Simple Machines

© sciencebod 2014

Preface

Machines enable us do work more easily; they have gradually become part of our existence. This book presents ideas on the workings of basic machines in very understandable manner. There are lots of numerical examples and exercises that have been used to help facilitate the students' quantitative understanding of the subject. So, read and enjoy!

SIMPLE MACHINES

1.0 Introduction

1

It will be extremely difficult for one to carry ten gallons of water by hand to travel a reasonable distance, but using a wheel barrow for example, it will be much easier.

Any device which enables work to be done easily and more conveniently is called a machine

1.1 Definition

2

A machine is a device by means of which an effort E, applied at one point can be used to overcome a load L, at some other points.

OR

A machine can also be said to be a device which enables a large weight or resistance to be overcome by a small effort.

1.2 Examples of machines

3

In physics, a machine is not necessarily a piece of complicated mechanism. The following are examples of machines: A wheel barrow, pliers, an inclined plane, pulley systems, the screw, the lever, car lifting jack, wheel and axle etc.

1.3 Terms used in discussing simple machines

| 4 |

Before discussing the working principles of simple machines, we need to define some terms that apply to the working of machines namely: Mechanical Advantage (M.A), Velocity Ratio (VR) and Efficiency (Eff).

1.3.1 Mechanical advantage (M.A) or Force Ratio (FR)

| 5 |

The force applied to a machine is called the Effort E, and the force of resistance to be overcome by the machine is called the load.

Mechanical advantage is therefore defined as the ratio of load to effort. It is given by mechanical advantage $= \dfrac{Load}{Effort}$ 1.0

Because mechanical advantage is the ratio of two forces load and effort, it is sometimes referred to as the <u>Force Ratio</u> (FR).

Mechanical advantage can then also be defined as
$$MA = \dfrac{Output\ Force}{Input\ Force} \qquad 1.1$$

1.3.2 Velocity Ratio.

| 6 |

This is defined as the ratio of the distance moved by the effort to load, in the same time interval, i.e.

$$V.R = \dfrac{\text{distance moved by effort (a)}}{\text{distance moved by load (l)}} \qquad 1.2$$

For an ideal machine, work done by the machine is equal to the work done on the machine. i.e.,

Load x distance moved by load = effort x distance moved by effort.

OR $\dfrac{\text{Load}}{\text{effort}} = \dfrac{\text{distance moved by effort}}{\text{distance moved by load}}$

Hence,

Mechanical Advantage = Velocity Ratio

for an ideal or perfect machine.

1.3.3 Efficiency (Eff)

The ratio of the useful work done by the machine to the total work put into the machine is called the efficiency (Eff) of the machine. It is usually expressed in percentage thus:

$$\text{Eff} = \dfrac{\text{Useful work done by the machine}}{\text{work put into the machine}} \times 100\% \qquad 1.3$$

Since work (W) is given by; W = force x distance

$$\text{Eff} = \dfrac{\text{Load (L)} \times \text{distance moved by load (l)}}{\text{Effort (E)} \times \text{distance moved by effort (e)}} \times 100\%$$

$$= \dfrac{L}{E} \times \dfrac{l}{e} \times 100\%$$

$$= \dfrac{L}{E} \div \dfrac{e}{l} \times 100\% \qquad = \dfrac{\frac{L}{E}}{\frac{e}{l}} \times 100\%$$

$$= \dfrac{Mechanical\ advantage}{Velocity\ ratio} \times 100\%$$

Therefore, $\quad \text{Eff} = \dfrac{MA}{VR} \times 100\% \qquad\qquad 1.4$

Is it possible to obtain 100% efficiency in a machine?

8

The answer is No!

In practical machines the efficiency is usually less than 100%. This is because of friction in the moving parts of the machine.

In such practical machines, part of the effort applied is used to overcome frictional forces which are always present. Thus the useful work done by the machine is less than the work done by the effort on the machine.

Therefore it is only a perfect or ideal machine that has 100% efficiency and they do not exist practically.

Let's start our illustrations with this JAMB question

9

A machine requires 1000J of work to raise a load of 500N through a vertical distance of 1.5m, calculate the efficiency of the machine
(A) 80%   (B) 75%   (C) 50%   (D) 33%

Solution

10

$$\text{Efficiency} = \frac{wor\ done\ by\ the\ machine}{work\ done\ on\ the\ machine} \times 100\%$$

But work done by machine = Load x Distance moved by load
= 500N x 1.5m = 750J

Work done on the machine = 1000J

$$\therefore \text{Eff} = \frac{750}{1000} \times 100\% = 75\%$$

Option B is correct!

Another JAMB question

| 11 |

A machine whose efficiency is 60% has a velocity ratio of 5. If a force of 500N is applied to lift a load p, what is the magnitude of p?

(A) 1500N   (B) 500N   (C) 4166N   (D) 750N

Solution

| 12 |

$$\text{Efficiency} = \frac{Mechanical\, advantage}{Velocity\, Ratio} \times 100\%$$

$$\therefore \text{Eff} = \frac{\frac{Load\,(p)}{Effort\,(E)}}{VR} \times 100\% \qquad = \frac{Load\,(P)}{Effort\,(E)} \times \frac{1}{VR} \times 100\%$$

$$60 = \frac{Load\,(p)}{500} \times \frac{1}{5} \times 100\%$$

Therefore, Load $p = \dfrac{60 \times 500 \times 5}{100} = 1500N$

And option A is correct!

1.4 Types of machines

| 13 |

Types of machine include: the lever, the pulley systems, the inclined plane, the wedge, the screw, the hydraulic press, the wheel and Axle and Gear wheels.

We'll treat them one after another.

1.4.1 The Lever

| 14 |

The lever is further classified as first order, second order and third order. This classification is based on the relative position of effort, load and fulcrum.

First order lever

15

In the first order lever; the fulcrum (F) or pivot is between the load and the effort.

Examples of such levers are: (i) the crowbar, a pair of scissors of pincers, claw hammer and pliers.

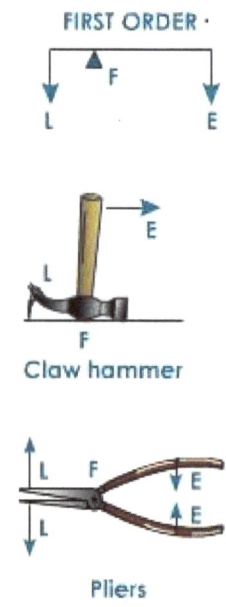

Figure 1. First order lever

Here, the velocity ratio is usually greater than 1 but it could be less or equal to 1.

Second order lever

16

Here, the load, L, is between the effort, E, and the fulcrum or pivot. Examples of this class of levers are: wheel barrows and nutcrackers.

Wheel barrow

Nut-cracker

Figure 2. Second order lever

In this case, mechanical advantage and velocity ratio are always greater than 1.

Third Order Lever

A third order lever has the Effort (E) between the load and the fulcrum F.

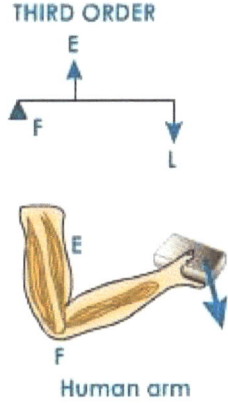

Human arm

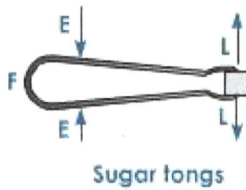

Sugar tongs

Figure 3. Third order lever

Examples of third order lever are: the forceps, tongs, forearm, etc.
Mechanical advantage and velocity ratio are always less than one 1.

Let's take a quick look at this JAMB question

18

Which of the following levers has the greatest mechanical advantage?

(A)

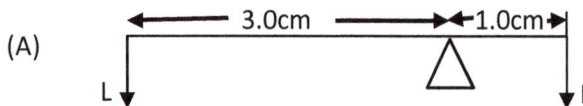

(B)

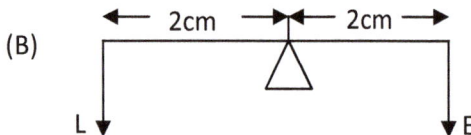

(C)

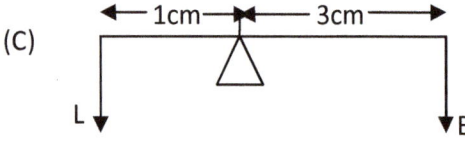

(D)

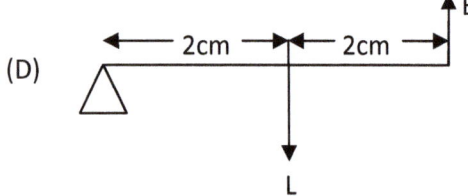

19

Since mechanical advantage $\propto$ V.R, we can compute indications of the mechanical advantage as follows:

$$\text{M.A} \propto \text{V.R} = \frac{\text{distance moved by effort}}{\text{distance moved by load}}$$

For lever (A): M.A $\propto \dfrac{1}{3}$

For lever (B): M.A $\propto \dfrac{2}{2}$

For lever (C): M.A $\propto \dfrac{3}{1}$

For lever (D): M.A $\propto \dfrac{4}{2}$

Of all the four, $\dfrac{3}{1}$ is the greatest, therefore lever (C) has the greatest mechanical advantage.

1.4.2 The Pulley systems

20

A simple pulley is a fixed wheel with a rope passing around a groove in its rim. A load L is attached at one end of the rope while the effort E is applied at the other end.

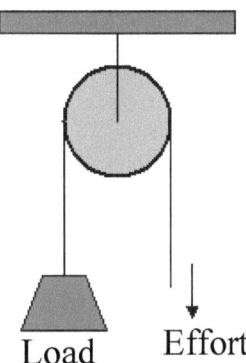

Load   Effort

21

If we neglect friction at the wheels and the weight of the rope, then tension T in the rope will be the same throughout. We will then have that at equilibrium of the system:

Load = Effort = Tension in the rope

Also M.A = V.R = 1

Since $M.A = \dfrac{Load}{Effort}$, and Load = Effort

It implies that M.A = 1.

Also since Efficiency $= \dfrac{M.A}{V.R} \times 100$, and Efficiency = 100% (Neglecting friction)

It implies that V.R = M.A

Therefore, for such systems, V.R = M.A = 1

Block and tackle system of pulleys.

23

A more practical system of pulleys is one in which more pulleys are mounted on the same axle, with one continuous rope passing all round the pulleys. This is called the block and tackle system of pulleys.

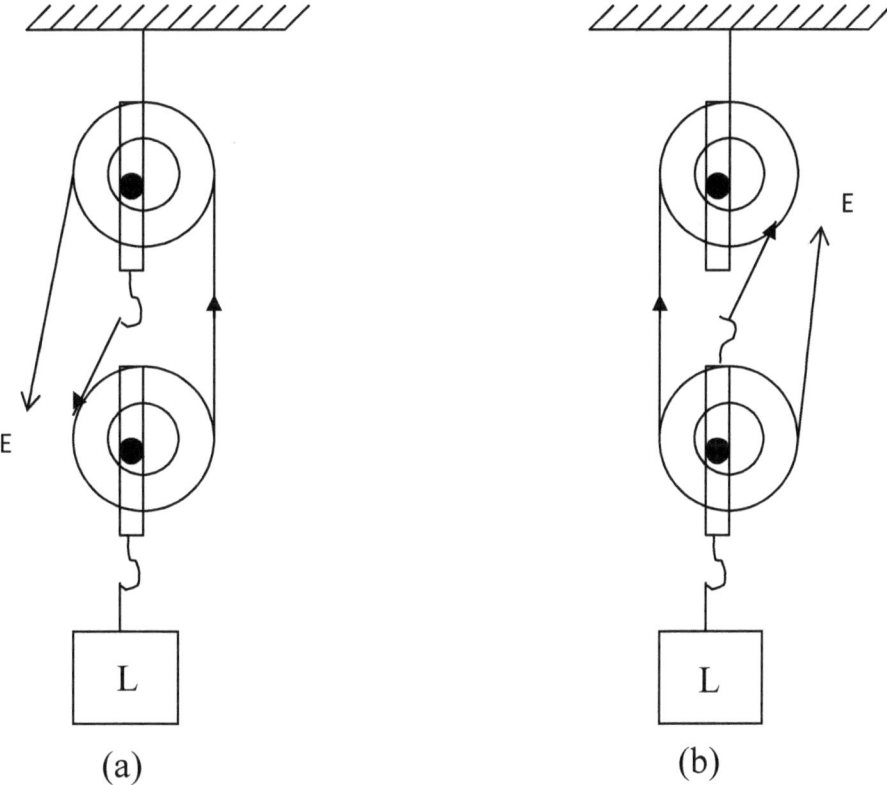

Figure 4. A Block and Tackle System of Pulleys.

In figure 4(a) and (b) above, the lower pulley is movable but the upper one is fixed. When an effort is applied at E, the wheels of the system rotate and the load is therefore pulled up.

V.R of a Block and Tackle System

24

The V.R of this sort of pulley system is generally equal to the number of the pulleys. It is therefore 2 in figures 4(a) and (b).

The M.A increases as the number of pulleys increases. If M.A is equal to V.R, then the efficiency of the system is 100%. But owing to friction and the weights of the pulleys, the efficiency is always less than 100%.

Efficiency usually gets less as the number of pulleys increases.

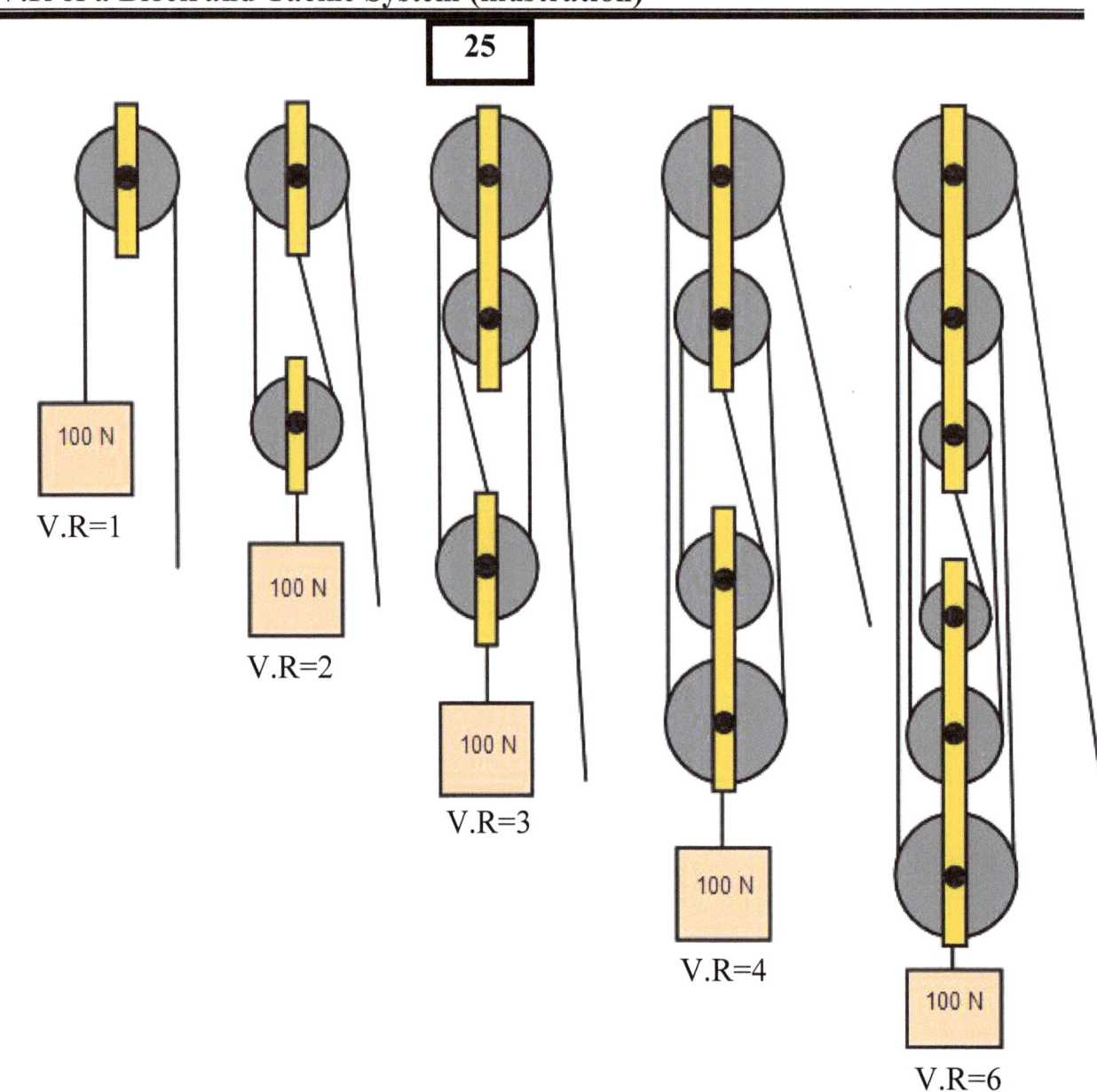

V.R=1

V.R=2

V.R=3

V.R=4

V.R=6

| 26 |

The diagram below illustrates a pulley system where an effort of 50N is applied to lift a load of 90N. Find the efficiency of the system.

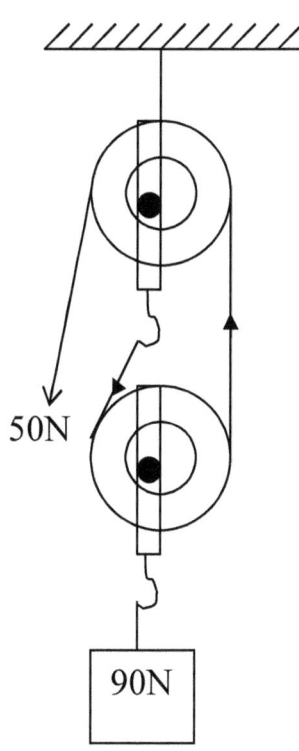

Solution!

| 27 |

We know from the foregoing that the velocity ratio of the system is V.R=2.

And the mechanical advantage is M.A = $\dfrac{Load}{Effort} = \dfrac{90}{50} = 1.8$

Therefore the efficiency is Eff = $\dfrac{M.A}{V.R} \times 100 = \dfrac{1.8}{2} \times 100 = 90\%$

And now this JAMB question!

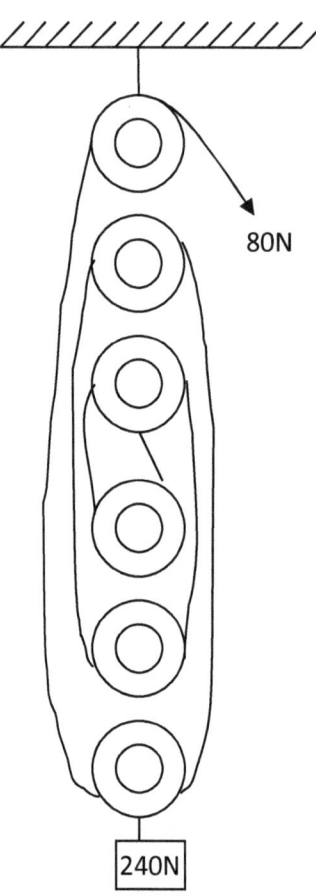

80N

240N

The diagram above is a block and tackle pulley system in which an effort of 80N is used to lift a load of 240N. The efficiency of the machine is
(A) 60%   (B) 50%   (C) 40%   (D) 33%

You should be able to get it right!

From the diagram, the velocity ratio V.R = Number of pulleys = 6

And the mechanical advantage is MA = $\dfrac{Load}{effort} = \dfrac{240}{80} = 3$

Therefore, the efficiency is Eff $= \dfrac{MA}{VR} \times 100 = \dfrac{3}{6} \times 100 = 50\%$

1.4.3 The inclined plane

30

The inclined plane is used to raise heavy loads such as drums of oil and engine blocks.

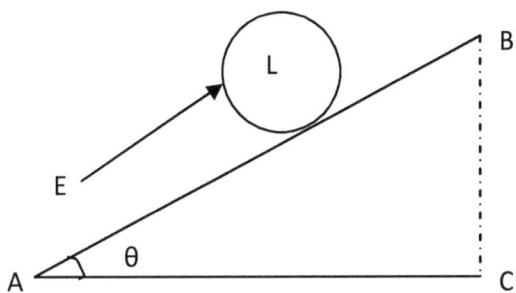

Figure 5. The inclined plane.

As shown in figure 5 above, a load L is moved by an effort E which moves a distance AB = X, the angle of inclination of the plane to the horizontal is θ. The load L moves a vertical distance h = BC.

The velocity ratio is V.R $= \dfrac{distance\ moved\ by\ Effort}{distance\ moved\ by\ Load} = \dfrac{AB}{BC} = \dfrac{X}{h}$

but $\sin\theta = \dfrac{BC}{AB} = \dfrac{h}{X}$

Therefore, V.R $= \dfrac{1}{\sin\theta}$ for an inclined plane

You should be able to answer this JAMB question

31

A plane inclined at angle θ has a velocity ratio of 10:1. The inclination of the plane to the horizontal is given by

(A) $\tan\theta = \dfrac{1}{10}$ (B) $\cot\theta = \dfrac{1}{10}$ (C) $\cos\theta = \dfrac{1}{10}$ (D) $\sin\theta = \dfrac{1}{10}$

Solution:

The velocity ratio is V.R = 10/1 = 10

And we know that for an inclined plane, the velocity ratio is V.R $= \dfrac{1}{sin\theta}$

$\Rightarrow 10 = \dfrac{1}{sin\theta}$ OR $sin\theta = \dfrac{1}{10}$

So, option D is correct!

1.4.4 The wedge

This is a small triangular block which is driven between two objects to force them apart. An example is a chisel.

When the wedge is driven in a distance X_1, the surfaces being separated are moved apart through a distance X_0 as shown in the diagram below.

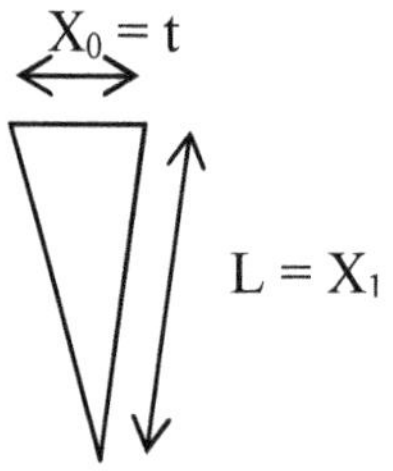

Figure 6. The Wedge

If the length of the wedge is L and its end thickness, t, then the mechanical advantage (neglecting frictional forces) is given by

$$M.A = \dfrac{Load}{Effort} = \dfrac{X_1}{X_0} = \dfrac{L}{t}$$

$$\therefore M.A = \dfrac{Slant\ height\ of\ the\ wedge}{thickness\ of\ the\ wedge.}$$

From the formula in the above plan, it follows that: a long, thin wedge has a higher mechanical advantage than a short thick one.

Or more generally, the smaller the angle θ between the slant heights, the grater the M.A

1.4.5 The screw

The screw can be thought of as an inclined plane wrapped round a cylinder to form a thread. The simplest example of a screw is a nut and bolt. As the nut is turned, it moves along the thread of the bolt as though traveling up an inclined plane.

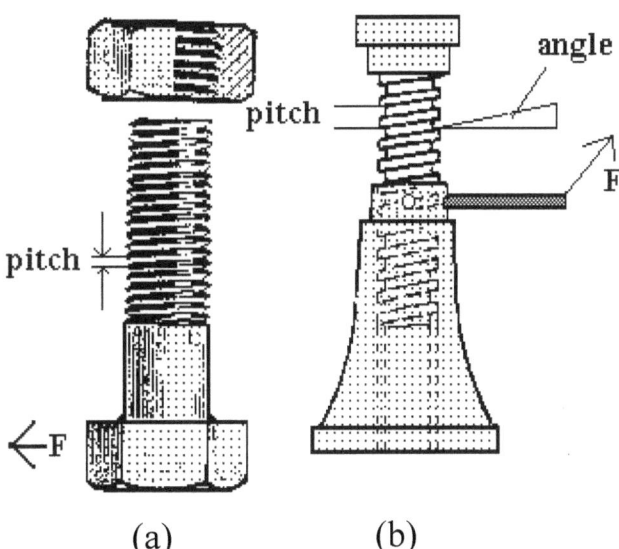

(a)   (b)

Figure 7. The screw jack

Things to know about the screw!

The distance between successive screw threads is called the pitch p.

When the screw head is turned through one complete revolution, the screw (load) moves toward through a distance equal to the pitch.

In the case of a screw jack as shown in figure 7(b), the effort is applied by means of a handle. If the handle (called 'the tommy bar') moves round once, the input force or effort acts through a distance equal to the circumference of a circle of radius r, where r is the length of the tommy bar.

At the same time, the load moves up through distance equal to the pitch p of the thread. Hence the velocity ratio (V.R) of the screw is given by

$$V.R = \frac{2\pi r}{p} \tag{1.5}$$

We illustrate with this JAMB question

Calculate the velocity ratio of a screw jack of pitch 0.3cm if the length of the tommy bar is 21cm.

(A) $\frac{1}{140}\pi$ (B) 14π (C) 17π (D) 140π

Solution

Velocity ratio of the screw jack is V.R = $\frac{2\pi r}{p}$

And we are given that p = 0.3cm and r = 21cm

$$\Rightarrow \text{V.R} = \frac{2\pi \times 21}{0.3} = 140\,\pi$$

And so option D is correct.

Now, attempt this WAEC question

38

A screw jack with a tommy bar of length 12cm is used to raise a car through a vertical height of 25cm by turning the tommy bar through 50 revolutions. Calculate the approximate velocity ratio of the Jack (Take $\pi = 3.14$)

(A) 21   (B) 38   (C) 48   (D) 151

Did you get it? Find out here!

39

Length of the bar, r = 12cm

Observe that the pitch of a screw is the vertical height when the tommy bar is turned through 1 revolution.

Now in this question, 50 revolutions of the tommy bar raised the car through a vertical height of 25cm. Therefore 1 revolution of the tommy bar will raise the car through a vertical height of 25cm/50 = 0.5cm
(This is the pitch of the screw)

$$\Rightarrow \text{V.R of the Jack} = \frac{2\pi r}{p} = \frac{2 \times 3.14 \times 12}{0.5}$$
$$= 150.72 \cong 151$$

And so option D is correct.

1.4.6 The hydraulic press

A press is a device used to produce a very large force, to compress a fluid. The figure below shows a typical hydraulic press.

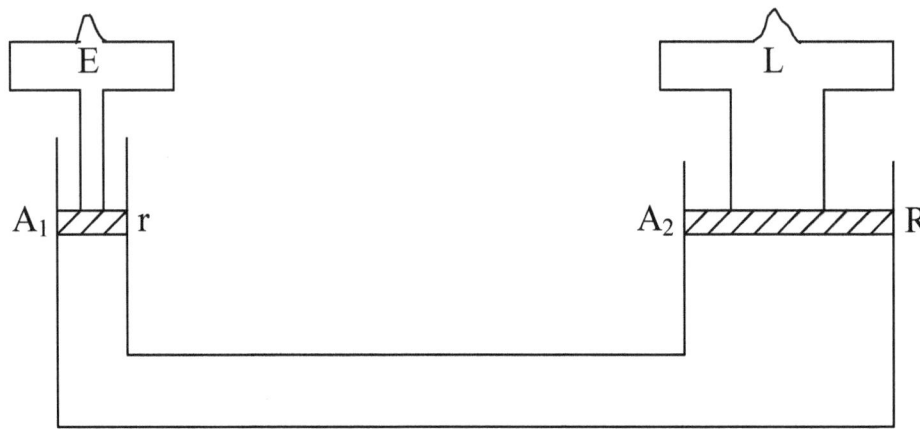

Figure 8. The Hydraulic Press

An effort (E) applied could lift a large load (L) as shown in the figure above.

If A_1 is the area of the small piston where an effort (E) is applied, by the definition of pressure we have that

$$P = \frac{E}{A_1} \qquad \text{OR} \qquad E = PA_1 \qquad\qquad 1.6$$

Similarly, if A_2 is the area of the large piston carrying the load (L), then

$$P = \frac{L}{A_{21}} \qquad \text{OR} \qquad L = PA_2 \qquad\qquad 1.7$$

Now the mechanical advantage is

$$\text{M.A} = \frac{Load\ L}{Effort\ E}$$

And replacing the values of L and E from equations 1.7 and 1.6, we get

$$\text{M.A} = \frac{L}{E} = \frac{PA_2}{PA_1} = \frac{A_2}{A_1} \qquad\qquad 1.8$$

If we assume that the efficiency of the system is 100% (neglecting friction), then M.A = V.R

$$\therefore V.R = \frac{A_2}{A_1} = \frac{\pi R^2}{\pi r^2} = \frac{R^2}{r^2}$$

$$V.R = \frac{A_2}{A_1} = \frac{R^2}{r^2} \qquad\qquad 1.9$$

The principle of a hydraulic press is that pressure is transmitted equally to all parts of a liquid at the same level.

Let's illustrate with this JAMB question

41

A hydraulic press has a large circular piston of radius 0.8m and a circular plunger of radius 0.2m. A force of 500N is exerted by the plunger. Find the force exerted on the piston.

(A) 8000N   (B) 4000N   (C) 2000N   (D) 31N

Solution

42

Pressure at the piston = pressure at the plunger

That is, $\dfrac{force\ exerted\ by\ plunger}{area\ of\ plunger} = \dfrac{force\ exerted\ on\ piston}{area\ of\ piston}$

Therefore, $force\ exerted\ on\ piston = \dfrac{force\ exerted\ by\ plunger \times area\ of\ piston}{area\ of\ plunger}$

$$= \frac{500 \times (\pi \times 0.8^2)}{(\pi \times 0.2^2)}$$

$$= 8000N$$

| 43 |

In a hydraulic press, a force of 40N is applied on the effort piston of Area 0.4m². If the force exerted on the load piston is 400N, the area of the large piston is

(A) 4m²   (B) 8m²   (C) 2m²   (D) 1m²

Solution

| 44 |

$$\frac{L}{E} = \frac{A_E}{A_L}$$

where A_E = Area of the effort piston and
A_L = Area of the large (load) piston

$$\Rightarrow \frac{40N}{400N} = \frac{0.4}{A_L}$$

$$\therefore A_L = \frac{400 \times 0.4}{40} = 4m^2$$

1.4.7 The wheel and Axle

| 45 |

The wheel and axle system consists of a round cylindrical wheel of which a rope is wound round, leaving a free end where the effort is to be applied.

Attached to this wheel is another cylinder, the axle, having a common axis with the wheel.

The load is tied to a rope wound round the axle in the opposite direction to that of the wheel as shown in the figure 9(b) below.

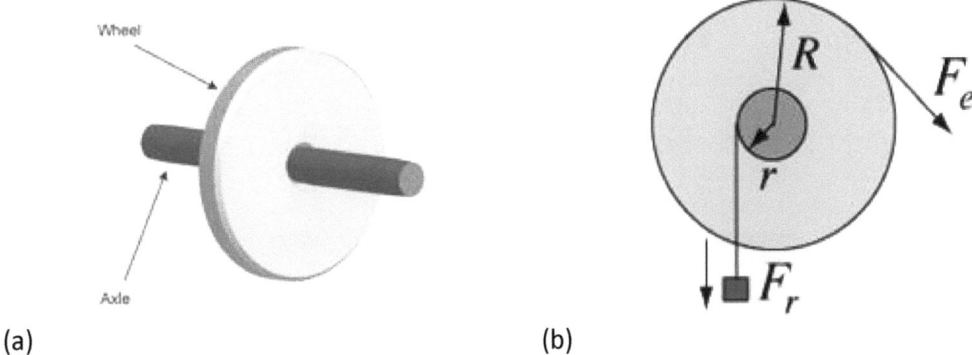

(a)   (b)

Figure 9. The Wheel and Axle

V.R of the wheel and axle system

46

When the wheel makes one complete rotation, the axle also makes a complete rotation. Therefore the effort, E moves a distance equal to the circumference of the wheel, while the load L moves a distance equal to the circumference of the axle.

If r and R are the radii of the axle and the wheel respectively, then the velocity ratio is:

$$V.R = \frac{distance\ moved\ by\ effort}{distance\ moved\ by\ load} = \frac{circumference\ of\ wheel}{circumference\ of\ axle}$$

$$\therefore V.R = \frac{2\pi R}{2\pi r} = \frac{R}{r} \qquad\qquad 1.10$$

The answer to this JAMB question should be evident

47

In an ideal wheel and axle system, R stands for the radius of the wheel and r is the radius of axle. The mechanical advantage is

(A) $\left(\dfrac{R}{r}\right)^2$ (B) $\dfrac{r}{R}$ (C) $\left(\dfrac{r}{R}\right)^2$ (D) $\dfrac{R}{r}$

Hint: For an ideal system (100% efficient system), the mechanical advantage is equal to the velocity ratio.

From the previous Plan, the correct option is: (D) $\dfrac{R}{r}$

Let's next look at this JAMB question

48

In a wheel and axle mechanism, the diameter of the wheel and axle is 40cm and 8cm respectively. Given that the machine is 80% efficient, what effort is required to lift a load of 100N?

(A) 20N   (B) 25N   (C) 50N   (D) 80N

Solution

49

Diameter of the wheel = 40cm   (∴ radius of wheel = 20cm)
Diameter of the axle = 8cm   (∴ radius of axle = 4cm)

$\Rightarrow$ Velocity ratio of the machine = $\dfrac{20}{4}$ = 5

Now, Efficiency Eff = $\dfrac{M.A}{V.R} \times 100$

And since $M.A = \dfrac{Load}{Effort}$

$\Rightarrow$ Eff = $\dfrac{Load}{Effort} \times \dfrac{1}{VR} \times 100$

$80 = \dfrac{100}{Effort} \times \dfrac{1}{5} \times 100$

$\therefore$ Effort = $\dfrac{100}{80} \times \dfrac{1}{5} \times 100$

= 25N

1.4.8 Gear wheels

Gears work on the same basic principles as the wheel and Axle. They are commonly used in cars, bicycles and cranes.

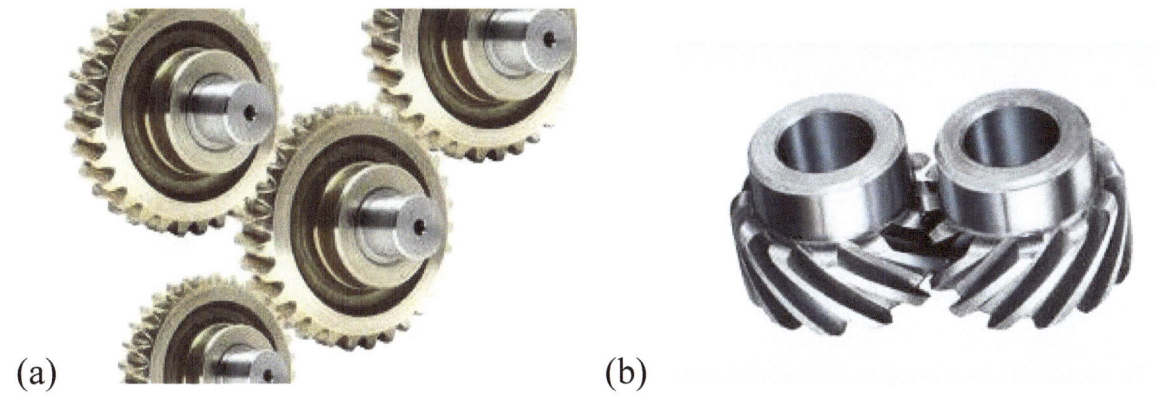

(a)   (b)

Figure 10. The Gear Wheels

Principle of gears:

Figures 10(a) and (b) above show typical gear wheels. Usually, one of the wheels is used to drive another wheel by means of their teeth connected.

The velocity ratio of the system is given by:

$$\text{V.R} = \frac{number\ of\ teeth\ on\ the\ driven\ wheel}{number\ of\ teeth\ on\ the\ driving\ wheel} \qquad 1.11$$

(1) A block and tackle system has 6 pulleys. If the efficiency of the machine is 60%, determine its mechanical advantage
(A) 12.0   (B) 10.0   (C) 3.6   (D) 1.8

(2) A simple machine with an efficiency of 75% lifts a load of 5000N when an effort of 500N is applied to it. Calculate the velocity ratio of the machine.
(A) 10.0   (B) 13.3   (C) 17.5   (D) 25.0

(3) A load of mass 12kg is raised vertically through a height of 2m in 30seconds by a machine whose efficiency is 100%. Calculate the power generated by the machine (Take $g = 10ms^{-2}$).
(A) 60N   (B) 80N   (C) 100N   (D) 120N

(4) A machine of velocity ratio 5 is used in lifting a load with an effort of 500N. If the machine is 80% efficient determine the magnitude of the load.
(A) 250N   (B) 2000N   (C) 900N   (D) 625N

(5) A machine of efficiency 80% is used to lift a box. If the effort applied by the machine is twice the weight of the box, calculate the velocity ratio of the machine.
(A) 0.50   (B) 0.63   (C) 0.80   (D) 1.60

(6) The efficiency of a machine is always less than 100% because the
(A) Work output is always greater than the work input
(B) Load lifted is always greater than the effort applied
(C) Effort applied is always greater than the load lifted
(D) Velocity ratio is always greater than the mechanical advantage

(7) A machine of velocity ratio 6 requires an effort of 400N to raise a load of 800N through 1m. Find the efficiency of the machine.
(A) 55.6%   (B) 50.0%   (C) 33.3%   (D) 22.2%.

(8) A machine is said to be a third order lever when the
(A) Load is between the fulcrum and effort
(B) Fulcrum is between the effort and load
(C) Effort is between the fulcrum and load
(D) Fulcrum is directly below the load

(9) Which of the following actions will improve the efficiency of a pulley system?
   (A) Reducing the mass of the pulley
   (B) Increasing the frictional force between the string and the pulley
   (C) Increasing the mass per unit length of the string of the pulley
   (D) Increasing the mass of the pulley.

(10) A block and tackle system has six pulleys. A force of 50N applied to it lifts a load of weight (w). If the efficiency of the system is 40%, calculate w
(A) 300N   (B) 200N   (C) 40N   (D) 120N

(11) The efficiency of the pulley system shown below is 80%. Find the effort required to lift a load of 1200N (A) 275N   (B) 325N   (C) 375N   (D) 1573N

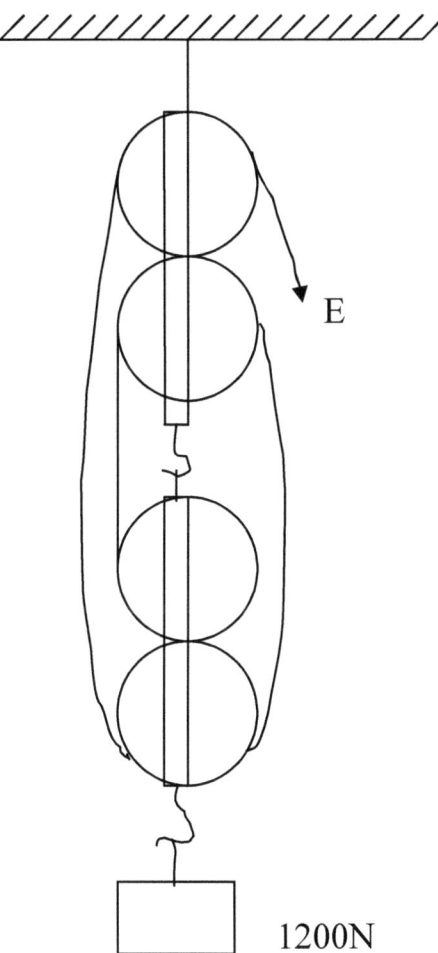

E

1200N

(12) A plane inclined at 30^0 to the horizontal has an efficiency of 50%. Calculate the force parallel to the plane required to push a load of 120N uniformly up the plane.

(A) 50.0N   (B) 120.0N   (C) 200.0N   (D) 240.0N

(13) If a heavy barrel is rolled up a plane inclined at 30^0 to the horizontal, its velocity ratio will be

(A) 3.0   (B) 3.1   (C) 3.2   (D) 2.0

(14) In a hydraulic press, a force of 40N is applied to the smaller piston of area 10 ms^{-2}. If the area of the larger piston is $200cm^2$. Calculate the force obtained.

(A) 800 N   (B) 500 N   (C) 80 N   (D) 50 N

(15) A screw Jack with 25% efficiency and having screw of pitch 0.4cm is used to raise a load through a certain height. If in the process, the handle turns through a circle of radius 40.0cm, calculate the
(i) velocity ratio of the machine
(ii) mechanical advantage of the machine
(iii) effort required to raise a load of 1000N with the machine (Take $\pi = 3.14$).

(16) An electric water pump rated 1.5kw, lifts 200kg of water through a vertical height of 6 meters in 10 seconds. What is the efficiency of the pump?
(Take g = $10ms^{-2}$, neglect air resistance)

(A) 90%   (B) 85%   (C) 80%   (D) 65%

1. C
2. B
3. B
4. B
5. B
6. D
7. C
8. C
9. A
10. D
11. C
12. B
13. D
14. A
15. (i) V.R = 628, (ii) M.A = 157, (iii) E = 6.4N
16. C

References

53

Anaheimautomation: www.anaheimautomation.com (Figure 10b)

Boston University: ed101.bu.edu (Figure 9a)

Glogster: http://www.glogster.com (Pulley image)

Hyperphysics: hyperphysics.phy-astr.gsu.edu (Figure 9b)

Photo-Dictionary: www.photo-dictionary.com (Figure 10a)

Proprofs: www.proprofs.com (Figures 1, 2 and 3)

SchoolPhysics: http://www.schoolphysics.co.uk (Block and Tackle image)

UQ: www.uq.edu.au (Figure 7)